I0639508

H. J. L. G.

Medleys and Songs without Music

H. J. L. G.

Medleys and Songs without Music

ISBN/EAN: 9783744729956

Printed in Europe, USA, Canada, Australia, Japan

Cover: Foto ©berggeist007 / pixelio.de

More available books at **www.hansebooks.com**

MEDLEYS

AND

SONGS WITHOUT MUSIC.

MEDLEYS

AND

SONGS WITHOUT MUSIC.

BY

H. J. L. G.

LONDON :
W. H. ALLEN & CO., LIMITED, WATERLOO PLACE.

1891.

LONDON:

PRINTED BY W. H. ALLEN AND CO., LIMITED,

13 WATERLOO PLACE, S.W.

CONTENTS.

ERRATA.

Page 102, line 16, *for* na mie *read* ma mie.

.. 104. .. 19, *for* concon *read* cancan.

.. 111. . 12, *for* Krankes *read* Kranke.

ROMEO AND JULIET.

ROMEO AND JULIET.

(Original Lancers.—JULLIEN.)

The story which I'm going to sing
Is one which will your soft hearts wring,
And to your needy optics bring
  A sympathetic ocean;
And if you're melted, pray mop your grief
In the silent silk of a handkerchief,
But don't be shrieking or seek relief
  In vulgar loud emotion!
For 't is a tale of dismal woe,
That happened many a year ago,
Of which, if you your Shakespeares know,
  You've probably got some notion.

("The Cork Leg.")

At Verona there dwelt two familees,
Who never each other at all could please;
There was always a row when both were present,
And it kept the dull city both lively and pleasant!
  Ri tooral, &c.

1 *

If a Capulet met a Montague,
Bad language would first as a rule ensue,
And then each would try his hands to imbrue
In the blood of the other—which blood it was blue—
          Ri tooral, &c.

The prince of the town would rush in at the noise
With his army—six old men and four little boys—
But this well-meaning, well-dressed potentate
Invariably arrived ten minutes too late!
          Ri tooral, &c.

("*A Many Years Ago.*"—PINAFORE.)
Now this was most affecting!
     However could they do it?
All decency neglecting,
     For every creature knew it.
However could they do it?
They never seemed to rue it,
But daily would renew it—
     That most unseemly fray!

Two tender children grew,
     Each had its private mother,
The one a Montague,
     A Capulet the other!
Now this was the position,
Each family was patrician,
And each in good condition
     From fighting every day!

("*È Scherzo, od è follia.*"—BALLO IN MASCHERA.)

Lord Capulet he gave a ball
  His neighbours to amuse,
And great and small he asked them all,
  Except the Montagues ;
But Romeo said, " I must go,
My heart does tell me so !
  And, as I wasn't asked,
  I'd better go all masked,
And toddle in incognito ! "

("*Tenting To-night.*"—CONFEDERATE SONG.)

I'll be dancing to-night at the old man's ball,
  Many are going there ;
Some wear high gowns, some quite low,
  Others cut them square.
Many are the feet that will weary to-night,
  Hopping on the well-waxed floor,
Many the mammas of large appetite
  Waiting at the supper door !
Dancing to-night ! dancing to-night !
  Prancing in the old man's hall !
Feeding to-night ! flirting to-night !
  Fooling at the old man's ball !

("*Just look at that.*"—CLOCHES DE CORNEVILLE.)

And old Capulet was standing
At the top of the first landing,
   All in his best,
   Greeting each guest.
And his heart was light and merry,
And they said it was the sherry
   That had made him so prime,
   So utterly sublime!

"A glance give here! a glance give there!"
  ('T was thus he pointed out the fair)
"Just look at that! just look at this!
  Now don't you think it's not amiss?
  A glance give there! a glance give here!"
He was a flirt—that aged peer!

(*Recitative—to Music from "Roméo é Giulietta."*)

   Then sudden with a start oh!
   And a feeling agitato at the heart oh!
     Roméo was gazing,
   All his soul from his eyes forth blazing!

("*Over There.*"—CHRISTY'S.)

"Quick tell me, Mercutio, true,
  Who is that pretty girl dressed in blue?
  I ne'er saw such a screamer, did you?
  And I feel an experience new.

Through and through, through and through,
An experience totally new,
Such as few young men do,
When they see pretty girls dressed in blue."

("*Let's give three Cheers for the Sailor's Bride.*"—PINAFORE.)

MERC. " Oh ! that is the lovely Miss Juliet,
The pride of the family Capulet,
And of all the girls in the neighbourhood
There's none that's so pretty and none so good ! "

(*Recitative—to Accompaniment of Valse in* " *Faust.*")

Then he spoke to her in Italian.
They all spoke Italian quite fluently at Verona !
" Ah ! non fuggir, mia bella Giulietta,
La ci darem la mano, mia petta,
Il mio tesoro,
Io t' amo, io t' adoro ! "
JUL. " Ah prego, non piu resta,
Son io figlia modesta,
E non voglio un matrimonio segreto ! "

(" *Permettereste a me.*"—FAUST.)

ROM. " Stay, pretty damsel, stay ; O do not run away oh !
Or you will disappoint most awfully your Roméo."

Jul. " No, signor, no, signor, I am single and unprotected,
  And I 'd rather—I would rather—
  If you would let me go and look for my mamma."

(" *Minuet.*"—Don Giovanni.)

 But Romeo would not leave his Juliet
 Till they had danced a minuet,
 Which mildest form of exercise
 Our ancestors did greatly prize ;
 Standing on tiptoe stately,
  Stepping the figures one by one,
 Bowing low and courtesying,
  That 's the way the dance is done !

(" *Old Lancers.*")

And then with some other good dancers too
They had a most capital Lancers too ;
They remembered the figures and danced when they ought,
And they did pretty steps, for their dresses were short.

(" *Bric-a-brac Polka.*")

 Then they danced the polka quite divinely
  (Each had had some lessons in the dance),
 And they did their one, two, three, hop, finely
  With a sort of pleasing, playful prance !

First they went the right way, then the wrong way,
    Up and down the crowded ball-room floor,
But it never seemed the least a long way,
    Till at last the dance was nearly o'er,

("*A Fine Old English Gentleman.*")

For midnight struck, and Capulet, who was an early bird,
Thought that bed could not with prudence any longer be
    deferred ;
So he called on Juliet, who was drinking teacups without
    number,
To clasp her bed-room candle and to seek her downy
    slumber,
    Like a good obedient maiden, all of the olden time !

("*A Game of Speculation.*")

    But she couldn't sleep a wink,
    She could only sit and think
At her window (which it overlooked the garden) ;
    For though usually blest
    With abundant powers of rest,
For her bed that night she didn't care a farden !

    " Ah me ! "
    To herself said she,
And she heaved a style of sigh quite undefeated !

So that Romeo, who was under,

Gazing up in silent wonder,

Dearly wished that that remark might be repeated !

("*There is a Young Woman.*")

Oh Romeo ! say, oh !

Why art thou Roméo ?

Disown your relations and let us be gay oh !
What 's in a name ?
What 's in a name ?

Called Smith or Tompkins, you'd smell
just the same !

("*He is an Englishman.*"—PINAFORE.)

For he is a Romeo !

He might have been an Othello,

Or some other Shakespearian fellow,

Macbeth, or Bassanio,

Shylock, Hamlet, or Banquo,

King Lear, or Desdemona,

Or a gentleman of Verona,

Or the Prince of Morocco ;

But, in spite of all temptation

To change his appellation,

He remains sweet Romeo !

("The Bailiff's Daughter of Islington.")

At the end of such a run
Juliet's vocal powers were done,
And Romeo, observing his chance,
Struck a lively bar
On his gay guitar,
And sang this chaste romance,
Interspersed with a breakdown dance.

("I wish I was a Mouse."—LITTLE DR. FAUST.)

"I wish I was a mouse,
A cockroach, or a spider,
I'd climb right up the house
And sit me down beside her!
I wish I was the kid
On which her cheek reposes,
Instead of being hid
Down here among the roses!
For I never yet my affection set
On a nicer pet, on a sweeter pet,
I never met a prettier pet
Than Miss July Capulet!

"I wish I was a bird!
The wish may seem absurd,
But I do, upon my word,
Desire to be a flapper!

That I might pipe and trill
Upon her window-sill,
And with my little bill
Most lovingly might tap her!
"For I never yet," &c.

(*Recitative.*)

Then Juliet, feigning to be much surprised,
Her lover thus lovingly apostrophized!

("*At it again.*"—LITTLE DR. FAUST.)

"At it again! At it again!
You've told your love once and you've told it me
    plain;
If papa took a shot,
He'd you probably pot,
And the dog in the yard has a very long chain!
Why do you do it? Why do you do it?
You'll catch a cold in your head;
Ain't it enough you to tell that I love you?
Why don't you go to bed?"

(" *Tom Bowling.*")

But Romeo was a person of superior courage,
   Nor dog nor man did he fear;
He asked her if she'd take him in marriage,
   And she said, "I will, my dear."
He much would have liked on the spot to elope,
   His heart was remarkably soft;
But, as he'd neither a ladder nor rope,
   He could not go aloft!

(" *My Mother had a Maid called Barbara.*")

Now Juliet had a nurse called Barbara!
   She was in bed, she was in bed,
For a cold she caught proved bad and wouldn't forsake
   her.
She had a sort of pillow—an old thing 't was—
But it expressed her passion, and she threw it at Juliet!
   " This cold to-night will not go from my head,
     Do shut that window and jump into bed!
I've much ado not to be blown out of bed all on one
   side!
   Such conduct's truly barbarous,
     It's barbarous!"

("*Grandfather's Clock*.")

Then she said "Bon soir," but she meant "Au revoir,"
  For they'd settled to meet in a trice;
And "the question," he said, "is to wed or not to wed?
  And the monk is the man for a splice."
'T was a nice pleasant parson, whom nothing ever
    shocked,
  And he oft-times had told him his flame;
So he knocked, knocked, hammer and tongs he knocked,
  Till the old man came.
        Slamming at the cell,
          With a bang, bang, bang, bang!
        Pulling at the bell,
          With a clang, clang, clang, clang!
        He thumped, swore,
        Thundered at the door,
          Till the old man came.

("*My Name is John Wellington Wells*."—SORCERER.)

*My name's Friar Lawrence, he said,
  And, if anyone wants to be wed,
    I'm ready to do it,
    Although they may rue it,
  When honeymoon fancies are fled.
There are rings by the gross or the pound,
And licenses here to be found,

   * This number was contributed by A. E. G. H.

If you want to play pranks,
You can fill up the blanks,
And they 'll do for your case I'll be bound.
We can find you in cake and in wine,
　　Country orders are punctually attended;
But so great an attraction is mine,
　　That the free list 's entirely suspended!
We 've broughams and postboys by dozens,
　　Also forms of proposal we keep,
And for bridesmaids we 've beautiful sisters and cousins,
　　And best men remarkably cheap!

　　　　Speeches we make for you,
　　　　Tidings we break for you,
　　　　Find you in merriment,
　　　　Try the experiment!
　　　　All of our jollity
　　　　Very best quality!
　　　　Fathers to give away,
　　　　Mothers to live away!
　　　　We 've Gunters attendant, too,
　　　　Waiters resplendent, too!
　　　　Always particular,
　　　　Quite perpendicular!
　　　　Every variety
　　　　Too of society,
　　　　And, if you want it, we
　　　　Make a reduction on taking a quantity!
　　　　　　Oh!
　　　　　My name 's, &c.

(*Mendelssohn's* " *Wedding March.*"—2nd part.)

Friar, Friar, my heart's on fire!
No banns, no license I desire,
In circumstances so peculiar
Wed me at once to my own sweet Julia!
　　Soon then their vows were plighted—
Friar, most accommodating,
Wouldn't keep the couple waiting—
　　In wedlock's bonds united,
Montague with Capuletta mating!

(" *Legend of the Bells.*"—CLOCHES DE CORNEVILLE.)

　Now the marriage ended,
　Bride and bridegroom splendid!
Feeling all the bliss which language cannot tell;
　Happiness enchanting!
　Nothing, nothing wanting!
Everything so far had gone extremely well.
　Ding dong, ding dong, ding dong,
　Ding dong, ding dong, dell!
If they'd only known it,
　They'd have rung the bell!
If they'd known of the marriage,
　They'd have rung the bell!

(" 'T is Many a Year, my old Friend John.")

Now, as Romeo wandered down the street,
　He met the usual brawl,
And saw at fiery Tybalt's feet
　A friend expiring fall!
" Tybalt," he cried, " thou art my foe!
　Thy life hath reached its tether,
For thou hast slain Mercutio,
　And we were boys together!
Very little boys, merry little boys,
　We both were boys together.
We shared our joys, our tips, our toys,
　When we were boys together!"

(" I don't want to Fight.")

" I don't want to fight, but
　By Jingo! if I do.
I've got my sword, I've got my wind
I've got some science too!"
　So up! with one good slash
　He settled Tybalt's hash,
Before that he could spell Constantinople!

(" Rothesay Bay.")

But the Prince did Romeo banish,
　Far from the banks of Po,

And bade him promptly vanish
    To Bath or Jericho!
So he rose betimes in the morning, and quietly sneaked
    away
To a temporary residence in the town of Man-tu-a.

("*Will he no come back again!*")

Will he no come back again?
Spoilt my tale would be, 't is plain,
If at Mantua he remain,
So he *must* come back again!

("*Barbara Allen.*")

Now, old Capulet had set his mind
    On a son-in-law called Paris,
Said he, "That knot I'll promptly bind;
    Next week my daughter marries!"

("*The Artless Thing.*"—MADAME FAVART.)

But Juliet thought 't was quite enough
    To have *one* wedding ring,
    She was an artless thing!
She abused her Pa in language rough,
    Having her little fling;

And quickly by wire
She sent for the Friar,
That holy old Friar!
"O Friar," said she, "come, rescue me
From this parental tyrannee;
For Paris and I, we could never agree,
I am such an artless thing!"

("*Simon the Cellarer.*")

But the Friar he gave her a potion to drink,
  And he said, "My dear Miss C.,
    If kindly you'll shake it,
    Before that you take it,
  You'll find it a cure will be;
  'T will suit your complaint to a T!
For first a soft languor will over you creep,
Then for several days you will steadily sleep,
And (I trust that you follow me?) stretched on your bed,
As a door-nail or herring you'll seem to be dead;
    Thus, Ho! Ho! Ho!
    To sleep you'll go!
But you'll wake up again in a week or so!"

(*Recitative—to music of* "*The Mascotte Legend.*")
    Then to her throttle
    She puts the bottle,
    And with emotion
    She sips the soporific potion!

And o'er her stealing
A wondrous spell;
She now is feeling
Very far from well!

("*Glou, glou*," *Duett.*—MASCOTTE.)

And in recumbent attitude
See her unconscious lie,
With her eyes in slumber glued, glued, glued,
And her breath a silent sigh!
Glued, glued, Ah!

("*Don't make a Noise, or else You'll Wake the Baby.*")

Don't make a noise, or else you'll wake the lady!
Don't make a noise, or else you'll wake Miss Ju!
Don't make a noise—such conduct's very shady—
In Mausoleums people never do!
Hush! Hush!

("*Jenny Jones.*")

Imagine the sorrow
They felt on the morrow,
When they found pretty Juliet wouldn't awake!
The howls of her nurse,
And her parents took worse,
When of kisses or pinches no notice she'd take!

The whole of Verona
Sat down to bemoan her
With tears far more briny than Tidman's sea salt;
They did not cremate her,
But took in great state her
To place in the Capulet family vault.

("*Dear Little Buttercup.*"—PINAFORE.)

Poor little Juliet! dear little Juliet!
All for their Juliet cry;
Dear little Juliet! poor little Juliet!
Juliet, never say die!

Things would have been better
If the old Friar's letter
To Romeo had reached that young spark,
But the postal arrangements
Were Gothic and strange ones,
And so he was left in the dark!

He thought she was dead,
So the people all said,
And he knew that he mustn't survive;
So in garments of gloom
He went down to the tomb,
And he never came from it alive!

Poor little, &c.

(" *Three Jolly Sailor Boys.*"—Marzials.)

Now Paris, who I've stated
    Was her husband designated,
Had but newly come to the dismal cemetree,
        Sighing, "Ah me oh!" and "Oh me oh!"
        When up comes Mr. Romeo—
As awkward a rencontre as could be!

    "Draw, vill'in!
I'm resolved upon your killin',
Will you fight, fight, fight?
        I'm for fighting in the key!"
So he stabbed him hard and oftin,
And they placed him in a coffin,
        And the undertaker pocketed his fee.

(*Patter Song.*—C. Matthews.)

But Romeo foreseeing
    The end of his being,
Had purchased a poisonous mixture, a mixture,
        From a chemist cadaverous,
        Who sold it from avarice,
And said it would prove quite a fixture, a fixture.

Then Juliet uprising,
Declared 'twas surprising
Her husband had been so mistaken, mistaken;
And with Romeo's knife,
Like a dutiful wife,
She cooked her own beautiful bacon, her bacon!

("*Lord Lovel.*")
Thus Romeo he
Was a *felo de se*,
And Juliet herself too did diddle;
While Paris he died
Of a weapon applied
To his soft and too sensitive middle-iddle-iddle.

("*The Ratcatcher's Daughter.*")
Now these dismal facts
Are told in five Acts
By another more famous Reporter.
But I've done my best
To have them compressed
In a compass a little bit shorter;
I've squeezed up tight
A ball and a fight,
A marriage, and a quadruple slaughter—
I refer to Tybalt, Romeo,
Paris, and old Capulet's daughter!

("*The Cork Leg.*")

Now on the day my story ends
The Caps and the Montys were made fast friends,
And they spent the night in feasting and laughter,
And they all lived happy for ever after !

Ri tooral, &c.

("*È Scherzo, od è follia.*"—BALLO IN MASCHERA.)

Now the lesson I would inculcate
From this lugubrious song,
Is that to indulge in rage or hate,
Like cat and dog, is wrong !

Peace is not nice
At any price,
But don't unduly quarrel ;
And when you fight,
Be sure you 're in the right !
Now, could you have a better Moral ?

THE
CORSICAN BROTHERS.

THE CORSICAN BROTHERS.

◆

ACT I.

(*Ghost Melody.*—CORSICAN BROTHERS.)

GRIM and gory
    Is my story,
Fights to frighten, ghosts to scare!
    Hanky-panky!
    Brothers Franchi,
Lean and lanky—ghastly pair!!

("*In the Gloaming.*"—LADY ARTHUR HILL.)

In the gloaming, O my darlings—
    (Pardon this familiar tone,
But my tale is so appalling
    That I'm glad I'm not alone,)—
When the disembodied spirit
    Calmly glides across your view,
'T would be best to link your fingers,
    Best for ghost, and best for you.

(*" Es ist bestimmt."*—MENDELSSOHN.)

O Corsica, fair Corsica,
How strange thy children's antics are!
　　　How tight and green
　　　　Their velveteen!
　　How odd their walk!
　　How queer their talk!
Their manners how extremely chalk
　　　And coarse they are
　　　In Corsica, fair Corsica!

(*" Poor Mary Anne."*)

There young Fabian with his mother
　　　Dwelt, fit as flea;
While the other twin, his brother,
　　　Went on the spree.
But oh! the widow's eye grew dewy,
When she thought of distant Louey,
Who in Paris played the *roué*;
　　　Sad dog was he!

(*" Bid me to live."*—HATTON.)

But 'twixt those youths—I tell you truths--
　　　There was a mystic bond;
And, if the one was in the blues,
　　　The other would despond.

Should either feel a pain or ache,
    A fever or a chill,
The other would some medicine take—
    A timely draught or pill.

(Recitative.)

But this similarity was not confined
To a sympathy of feeling and of mind.
    For in body too
  They were similarly designed,
    And more like they grew
Both in front and behind.
      There were few
      Ever knew
      Which of the two
      Was Fabian and which Loo ;
      And they grew
      (As people do),

(" You grow more like your Dad every day.")

They grew more like each other each day,
They grew more like two peas every way ;
  Both in form and character
  So like, that one actor
Both brothers could easily play.

("*M' appari tutt' amor.*"—MARTA.)

   'T was on a night,
   Stage-moon shining bright,
     Fabian at
     Supper sat
   Without appetite;
He 'd only eat
Soup and fish and meat,
Some *entrées*, a relish, and a sweet.
  He was sad, and a sigh
'T was in vain he tried to smother,
  For most sharp is the eye
Of that creature called a mother.
   "O tell me the worst,
   Or my feeling heart will burst!
O say, Is there anything wrong?
Dis-moi dong, mong Fabiong!"

("*Ohé Mamma.*"—TOSTI.)

"O mamma, I to-day have felt
  Sensations most distressin';
I fear that our dear Louis may
  Have got some horrid mess in.
I feel a pain—a spasm—
I know dear Louis has 'em—
      Ohé mamma!

" Dear grandpa's clock, which used to go
   Most punctual to a minute,
 In half an hour is six weeks slow.
   I'm sure there's something in it.
 I wind, and wind, and wind it,
 It never seems to mind it—
                    Ohé mamma!

(Ghost Melody.)

   At this crisis,
   Up there rises,
 Through the floor a phantom grim.
   In a white shirt!
   (P'raps his night-shirt?)
 Is it Louis? Yes, 'tis *Him!*

(" *The Standard-bearer.*"—KÜCKEN.)

But ha! that stab that glistens on his shirt!
'T·is surely not a dab of ordinary dirt!
Ah, no! ah, woe! By everything that's holy,
It is his life-blood which is ebbing slowly!
The phantom, see! is pointing to the scenes which down-
     ward fall,
   (For all concerned the situation's trying.)

They gaze, see it all,
Through the gap in the wall—
Chateau-Renaud his blood-stained sword is drying,
And Louis at his feet, dear Louis, dying!

(*" Poor Uncle Ned."*)

In the poor old forest the morning sun is red,
In the forest of famed Fontainebleau;
And 't is there that Louis has laid his woolly head.
In the place where the grass ought to grow.
Hang all duelling, it 's low!
And hang that vile wretch Chateau-Renaud!
For poor brother L.
Has gone to the — well,
He has gone where all niggers *ought* to go.

(*Recitative.*)

Then cried Fabian,
Though to ride he was not able.
" Quick, saddle my Arabian!
'T is the best geegee in my stable!"

("*The Yeoman's Wedding-day.*"—Prince Poniatowsky.)

"Quick, bring me a hack!
 Don't clutch my coat-tails, mother;
I'll stick to its back,
 And thoughts of cowardice smother;
Though not much good at a horse,
 I'll cling to its *mane* by *main* force,
For I, as you know, am a Cors—
 Am a Cors-ican Brother."

("*My Love has gone a-sailing.*"—Molloy.)

But the neighbours, pitying, smile, and
 The geography explain,
That Corsica's an island
 Surrounded by the main;
His quickest way and surest
 To gain the Continong,
Is to ship as a Cook's tourist,
 For Marseilles or Toulong:
So Fabian's gone a-sailing,
 Sailing far away;
And I fear he's sadly ailing,
 To *mal de mer* a prey,
Ailing and pale-ing and wailing o'er the sea.
 O ship, sail quick!
 Dear Fabian's sick,
 A sailor bad is he;

O ship, sail sure!
He's really poor-
Ly, and longs on shore to be.

ACT II.

(*Recitative.*)

Dropping o'er Fabian, growing green and yellow
  A gorgeous most expensive velvet curtain,
Let us to Paris, where that other fellow,
  Louis, the fashionable twin, is flirtin'.

("*Auld Robin Gray.*")

Young Louis loved Emilia,
  And many more beside;
But nothing could be sillier,
  For she was another's bride;
She'd been already mated
  To the Admiral de Lesparre,
A somewhat antiquated,
  But benevolent old tar.
He had gone, that ancient mariner,
  On his watery avocation;
And p'raps he was playing the fool somewhere
  In a Naval Demonstration!
     But, Emily, feeling free,
     Was blithe as bird or bee,
And old Robins and young Robins came courting of she.

("*Polly Bluck.*")

She went to a ball, which was *masqué*, at the Opera,
O 't was improper, ah!
No one to stop her, ah!
She was just the sort of person who was sure to go a
cropper, ah!
A beauty by profession, P.B.
Now I don't object to gaiety, while youth and beauty
last,
But I *do* object to anything that verges on the fast;
And a ball with dominoes,
Is a thing I should oppose,
If Emilia belonged to me.
At the ball, bad luck!
She met that buck,
The wicked Chateau-Renaud, who at nothing stuck.
Ah! goodness gracious me!
He offered her a cup of tea!

("*Valse Song.*"—OLIVETTE.)

Then he, dancer renown'd,
Asked her, "Give me a round!
Nay, nay! be not so stern,
Just one, just one little turn;
'T is that sweet Olivette,
Of all valses the pet,

To so tripping a measure
Valsing's a pleasure
You ne'er can regret."
 And she knew,
 As she flew
Through the air, light and free,
 There were few
 Who could do
Such a valse-step as he!

Bⴟᴛ

("*Here's a Health to all Good Lasses.*")

Chateau-Renaud had a wager
That to supper he'd engage her,
To a meal champagney and oysterous,
With his boon-companions boisterous,
 'T was a vulgar sort of feast,
 (Chateau was a vulgar beast).
But young Louis, kindly interposin',
Stuck his most unlucky little nose in,
Said, "A friend of mine this lass is,
Let her pass, I charge you, asses!
 Fair Emilia, fear no harm!
 Feel the biceps on my arm!"

("*The Keel row.*")

Then Chateau, quickly flaring,
With rage mad stark staring,
A-swearing and a-tearing
   The beard upon his chin!
O wasn't this a real row,
A real row, a real row,
O wasn't this a real row
   That our dear L. was in.

("*Under the Willow.*"—CHRISTY.)

Under the oak-tree they fought with care,
   Chateau and Louis dei Franchi,
In his pocket a locket of golden hair,
   And the poems of Moody and Sankey!
Fair, fair, the fight was fair,
   Seconds and doctors attended;
But Louis was struck through the gizzard, where
   His gizzard could never be mended.

("*The Perfect Cure.*")

"You can't," says Gray,
"Do what you may,
   Recall the fleeting dust,
     Although you may turn
     To a storied urn,
Or an animated bust!"

You may try German baths,
Or Homœopaths,
Or anything else that's newer,
But you never can mend,
When a quarrelsome friend
Has stuck through your body a skewer.
A skewer, a skewer, a skewer, a skewer,
Of this now pray be sure,
When you've got a hole wide
In your little inside,
You must not expect a cure.

("*So early in the Morning.*")

But, ere he breathed his final sparks,
He made these *cursory* remarks,
As Chateau-Renaud, slightly bored,
Was wiping his ensanguined sword:
"I'm a ghost, but you'll soon be another,
For I'm off to tell my mother,
Ditto my *dittle* brother,
Before the break of day."

(Ghost Melody.)

But the scenery,
By machinery,
Once more opening, saves all trouble;
Fabian therein
Wildly glarin'
At the murderer of his double!

("Ring, ring the Banjo.")

Drop, drop the curtain,
I like that shade of red;
We're getting on, that's certain,
Thank goodness *one* twin's dead!

ACT III.

("*Il a l'oreille basse.*"—LE PETIT DUC.)

Chateau no longer tarries,
    Now his fell work is done,
He's off again to Paris,
    Fast as his legs can run.
      And though before
      He, by the score,
Had dipped his hands in his fellow-creatures' gore,
      He feels this time
      Remorse for crime,
A new and sore sensation at his core!
    And he resolves to mizzle,
      Quietly at dawn to go;
    Thus p'raps the Fates he'll chisel,
      Sly little fox, Renaud!
He must be quick to cut his stick!

("*The Forsaken.*"—V. GABRIEL.)

But he will return, 't is all my eye,
He cannot leave, in vain he'll try;
He *must* return, I'll tell you why,
He's doomed, you know, at Fontainebleau
      To die—to die.

(*" Es war ein Traum."*—Lassen.)

Then he and Montgiron, without more fuss.
Their clothes into portmanteaus cram ;
Then hail they a passing omnibus,
    Es war ein " Tram " !

(*" Tannhäuser March."*—Wagner.)

But Fate cannot so easily be cheated,
Destiny's a thing that will not be defeated ;
    What can't be cured
    Must be endured,
So I've from childhood been by copy-books assured.
    And, as they jog along,
    Beguiling time with song,
    And jests that doubtful sounded,
    There comes a crash !
    Something has gone wrong,
    An awful smash !
They're grounded and confounded,
And completely upside-downded !
    And they know it,
    It is Fontaine—*blow it,*
    They never seem to go it
Beyond the wood they hate so ;
    No use to fuss,
    No use to cuss,
    They from the omnibus
Must all descend, and calmly wait so.

("*The Harmonious Blacksmith.*"—HANDEL.)

Woodman! woodman!
Come there's a good man,
Fetch me a wheelwright, if you any know;
Is there a crack smith,
Some harmonious blacksmith,
Dwelling in the neighbourhood of Fontainebleau?
We were bound to Toulouse,
When a sudden screw loose
Tilted up our vehicle and lodged us in the snow,
Where we must continue,
Till some man of sinew
From the smithy come, so prithee don't be slow!

("*Over the Sea.*")

But who is that bloke
Wrapped in a cloak,
Striking an attitude under the oak?
Oh, does he thus pose
To frighten the crows?
Or is it some practical joke?
No, 'tis He, He, He,
We left with his face in
A basin,
Grimacin',
'Tis He, He, He,
From the Mediterranean Sea!

And he's travelled by day, and he's travelled by night.
He's not much to say, but he's ready to fight!
    His object you'll guess?
    What! homicide? Yes;
    But oh! justifiable quite.

(Recitative.)

Then Chateau Fabian sees,
And it makes his life-blood freeze!
"Ha! ha!—likewise Ho! ho!
That face I ought to know;
Those legs so strange, that lingo.
I recognize, by Jingo!
'T is He—or (as we say in French)
    C'est L(o)ui(s)."

(" What would you do, Love?"—LOVER.)

CHATEAU. How do you do, Sir?
    Pray, who are you, Sir?
    P'raps number two, Sir,
      From Corsicay?
    Or are you a vision,
    An apparition
    In lean condition,
      What want you, say?

FABIAN. I've crossed the ocean,
  In deep devotion,
  And 't is its motion
   That makes me blue;
  I've sought you—rot you!
  And now I've got you
  I mean to pot you,
   That's what I'll do.

("*The Matron of an Hour.*"—OLIVETTE.)

CHATEAU. I gather what you say, but I don't feel in-
  clined
To give my passions play, and to murder all mankind;
What would your mother do, if of both twins bereft?
Poor old bird, not one little chick-a-biddy left!
 No, no! no, no! That would be low!
 I'm much too generous to treat you so.
 Oh, dear me, oh! You little know
 The noble nature of Chateau Renaud.

("*Here's a first-rate Opportunity.*"—PIRATES OF PENZANCE.)

FABIAN. Oh, confound your magnanimity!
  You had best now sing " Nunc Dimitte ";
  For you can't expect impunity
  Now I've got my opportunity;

I have brought here weapons various,
So your life must be precarious,
You to stick without remorse I can
For such behaviour is strictly Corsican.

(*Recitative.*)

Then the two foes,
Removing some clothes,
In shirt and in hose
Most gracefully pose,
And at it they goes!

("*The King of the Cannibal Islands.*")

Then Chateau's attack was sharp and fierce,
He was parried in *carte*, he was parried in *tierce*.
No blow he could aim the guard could pierce
　　Of the man from the Corsican Island.
At length his sword was shivered in two,
You'd think they would hardly the fight renew?
You're *wrong*: stern Fabian still insists
On tying the sword-points to their wrists;
And so the battle begins again,
I can't the details of each round explain,
Suffice it to say that Chateau was slain
　　By the man from the Corsican Island.

(*Farandole.*—OLIVETTE.)

Ah! ah! ah! ah!
Sure the ghost will now slide in!
What is the use of waiting further,
Nobody else to murther?
Ah! ah! ah! ah!
Yes, the bogey will glide in,
If we'll only play again
That much too familiar strain.

(*Ghost Melody.*)

To the music
That makes you sick,
See, the ghost is rising slow!
For the last time
Takes his pastime
'Neath the glades of Fontainebleau.

Ghost unnerving,
Ghost unswerving,
Ghost by Irving,
Glide along!
Spectre rummy,
Clasp thy dummy!
Good night, Mummy!
Shut up, Song!

(Curtain.)

CINDERELLA.

CINDERELLA.

("*Rousseau's Dream.*")

LONG, long ago,
　　No matter what time,
Down in a kitchen dwelt a little maid;
　　And oh! dear, oh!
　　She had a hot time,
As ladies do when cooking is their trade.
　　Stepmother had she,
　　Ugly and bad she,
(Step-mas in stories never come out nice!)
　　So down in the kitchen
　　This bewitchin'
Maiden was left with the cockroaches and mice!

("*Phyllis is My only Joy.*")

But she'd step-sisters crabbed and rough,
　　Ancient as the hills or seas,
One would scold and one would cuff,
　　And they never failed to tease.

4

If anything went wrong,
They said 't was all along
Of Cinderella,
" Drat that girl, ah !
She must mind her Q's and P's!"
Cooking was her daily toil,
Drudging on her hands and knees,
Sometimes roast, and sometimes boil !
Yet she never seemed to please !

(" *Another jolly Row Downstairs.*")

They would beat her black and blue,
They would kick and scratch her, too ;
They would bang her with the tables and the chairs !
When the parents heard the clatter,
They only said " no matter !
There 's another jolly row downstairs ! "

(" *The Kerry Dances.*"—MOLLOY.)

Now, my hero, a prince so smart he,
Full of frolic and fancy-free,
Issued cards for an evening party,
" Dancing early, R.S.V.P.!"

Painting, powdering, puffing, rouging,
  Smoothing wrinkles each sister see,
Tight her waist and tighter scrooging,
  For the Prince was a great *parti*!
Oh! to think of it, oh! to dream of it
  Filled their hearts with glee!
O, the patented dress-improver!
  O, the fringe and the curling tong!
O, the feminine arts that move a
  Prince!—they laid them on hot and strong

(*"An Old Woman's Dream."*—MADAME FAVART.)

But Cinderella, sad and solitary,
  Sat by the fire;
Her hair, which was her own, was matted,
  And in rags her attire.
    She had no soap,
She never, never knew that form of piety;
    She had no hope
Of ever, ever being in Society,
  Of changing beetles and blacks
  For all the joys of Almack's!
But as she sat she saw a vision,
  'T was the ball and she was there.
The dress she wore was quite Parisian,
  And jewels sparkled in her hair!

And half a dozen youths advancing,
 Eldest sons and rich and tall,
With her were eager to be dancing,
 Alas! she could not wed them all!
Ah me! ah me! 't is but a dream!
And down the tears slowly stream!

  ("*Cheval de Bronze.*")

But see! by the dresser,
Expressly to bless her,
A figure (you'll guess her)
 Who wasn't before,
In costume so airy,
It must be a fairy,
Who for a vagary
 Has taken the floor.
Yes, that is the cause
Of her spangles and gauze,
Her skips so terrific, so supple her joints!
 She 's rather décolletée,
 And what is more faulty,
Her foot to the roof she persistently points!

  ("*Over the Garden Wall.*")

How did she get in here?
Isn't her conduct queer?
For mother would bolt the door and lock it,
And place the key in her under pocket;

How did she enter the mansion at all?
Why (probably) over the garden wall!

("*Student's Chorus.*"—FAUST.)

I've come from fairy land,
   God-mother true,
All your grief I understand,
   I'll pull you through!
You to the ball shall go,
   Do not despond!
Only let me touch you so
   With my little wand!
Soon her rags are goin',
   Disappearin',
Pretty garments take their place!
Oh what beauty she is shewin'
   Her novel gear in,
Blaze of diamonds and pearls and lace!

("*Old Men's Chorus.*"—FAUST.)

Ha! Ha! Ha! Ha!
   Isn't it nice?
   Horses from mice
   Changed in a trice!
And a coach by wondrous device
Turned from a pumpkin large and yellah!

Footmen by wizard's
Art formed from lizards,
Coachman in ermine
From a rat, big bloated vermin!
Ha! Ha! Ha! Ha! Ha! Ha!
Ain't she a swell ah.
Sweet maid-of-all-work, Cinderella?

(" *O dem Golden Slippers.*"—CHRISTY.)

And O dem glassy slippers!
O dem glassy slippers!
Slippers of glass de Fairy 's brought her
For her fairy feet!
O dem glassy slippers!
O dem glassy slippers!
Slippers of glass she 's gwine to wear,
Because dey are so neat.

(" *Two little wooden Shoes.*"—MOLLOY.)

And though if shoes you were buying,
It is not glass you would choose,
They 're pr'aps for the feet not more trying
Than two little wooden shoes!

*(" Prayer."—*ZAMPA.)

FAIRY. But now I would like
　　To give one little warning ;
　　Ere twelve doth strike
　　And night is turned to morning,

(" Never come back no more.")

　　Mind you come home once more,
　　Or at the palace door
Lizards, and rats, and mice, and pumpkins,
　　All will be as before !

*(" Prayer."—*ZAMPA.)

If past midnight your footstep lags,
All your dress will be turned to rags !

(Recitative—to Music of " Zampa " Overture.)

Cinderella rapidly is galloping towards her destination !
All along the route she is the subject of a flattering
　　ovation !
　　Soon the palace gates are past
　　(The horses, as I said, were fast),
　　They take the lady right in ;
　　Soon she is alightin',
Where the pampered menials are waiting to receive her
　　and to
Usher her with ceremony right into the presence of the

ROYAL FAMILY !

(" *Il destin.*"—CHORUS, HUGUENOTS.)

> See her now
> Gracious how
Fair she is as Hebe!
> All exclaim
> " What's her name?
Who on earth can she be? "

(" *Oh, She's charming.*"—MASCOTTE.)

" Oh, she's lovely! oh, she's a oner!
> Peerless, other maidens above! "
The Prince he thinks her quite a stunner,
> With her at once he falls in love!
" She's a oner! oh, she's a stunner!
> A real jam, a turtle dove! "

(" *Suoni la tromba.*"—TROVATORE.)

> Then he steps up to meet her,
> With courtly bow to greet her,
> Sure no vision ever sweeter
Caused a Prince such palpitation!
> At her he gazes,
> Until her eyes she raises
> To his face, which now betrays his
> > > Admiration.

Her little hand it lingers
In the middle of his fingers,
And he presses and caresses it in manner odd but
pleasant,
And he sings with delight
Scraps of Lytton and Maud White,
Which quite explain how tho' he's absent yet
he's present.

("*Absent yet present.*"—M. V. WHITE.)

As the fish in the river
That rise to the fly,
My soul rushing upward
Is hooked by thine eye;
It is not thy beauty,
Tho' that be displayed,
It is not my duty
To dance with each maid!
Look up! thou art near me,
And beauteous my face!
Look up! dost thou hear me?
And feast on each grace!
This is not the time and this is not the place
To clasp thee, to ask thee to feel my embrace!

(*Recitative.*)

But why this long delay while time is winging?
A ball's a place for dancing not for singing.

And see! the Royal band
    Awaits the Royal pleasure,
They ready stand
    To strike up any measure!
The Prince he thinks himself a lucky fellah,
For he has got the hand of Cinderella;
    And the sisters disappointed,
    With their noses quite disjointed,
    Wreathe in smiles their angry faces—
    Take their partners and their places!

(" *The Holly-bush Polka.*")

Then they polka so neatly,
For the band plays so sweetly,
And they know so completely
    What the dance ought to be!
To the right deftly steering,
To the left sudden veering,
All the while in time careering
    To the one, two, three!

(" *Oh, I love to think of the Days when I was Young.*")

Then they sat out several dances, which was nice!
And he took her into supper once or twice!
    She was hungry, so he collared
    Half a chicken and then holloaed
For a bottle of champagne in ice!

(" *Wise Folks have always Noted.*"—MASCOTTE.)

Next then a valse they 're trying,
    Ecstatic the pleasure,
Each a good dancer appearing!
Round, round the room they 're flying,
    Exchanging in measure
Question and answer endearing;
And as they whirl in giddy circles round the Hall
    Baronial,
He, feeling his affection for her growing more and more,
States that his aims and objects are entirely matrimonial,
    Will she be his for ever, his to cherish and adore?

(*Valse.*—Op. 34, No. 1.—CHOPIN.)

    " Say, say,
Dear, do you love me?"
    " Yea! yea!
By those gas-lights above me!"
" A Prince 't is his knees on
    Who humbly thus woos you!"
" And that is the reason
    I cannot refuse you!"
Then they treat the matter as completed,
In a pretty alcove snugly seated,
And their vows of love again repeated
    When Ha!!

(*Recitative.*)

She sees with shock,
The Royal clock,
Whose undefeated hand,
Doth at the hour,
(O gracious power!)
Of midnight nearly stand!
She starts up with horror!
It's almost to-morrer!
She must away,
She cannot stay,
With a hop and a skip,
And a jump and a trip,
While the guests and her lover, who wildly all stare
about,
Cannot discover the least bit her whereabout,
She's off! she's off! she's off!
Tho' she knows it will grieve him
To leave him!

("*Oh give Me back.*"—Mascotte.)

"Give me back my sweet Cinderella!
Give me back my poor broken heart!
Give me back my old umbrella!
For the tears in torrents will start
Give me back my Cinderella,
My umbrella,
And my broken heart!"

(*" Good-bye."* — Tosti.)

" Good bye, Cinderella,
    Good bye, good bye !
        Why fly,
        Mia bella ?
    Fie ! fie ! fie ! "

(*" The Minstrel Boy."*)

But our maiden coy to her home has gone,
    In her rags once more you will find her ;
No fuzzly fringe has she girded on,
    Nor a dress-improver behind her !

(*" She wore a Wreath of Roses."*)

She has no wreath of roses,
    As maidens always wear ;
No housemaid's cap reposes
    On her tossed and tangled hair.

(*" A Twopenny Ride in a Tramcar."*)

Fivepence she gave for her body,
    Fourpence she gave for her skirt,
Threepence she gave for her apron
    To concentrate the dirt !

Twopence she gave for her stocking,
  And a penny she hasn't to spare,
    For her bootmaker's bill
    Will be probably nil,
  As one of her feet is bare !

(*Mazurka.*—2nd *Set, Op.* 7.—CHOPIN.)

    You say how so ?
    Why don't you know
      That
     In her scare
    What did she do ?
  On the palace stair
  Dropt her tiny shoe,
    She left it there
    As off she flew ;
  The Prince who in despair
  Pursues the lady fair
    Exclaims " Hooroo !
    I have a clue !
I 'll find that lady by her little shoe ! "

(" *The Attractive Girl.*"—MASCOTTE.)

Then the criers, ancient fellows,
  All about the city ply,
Cracked their bells and cracked their bellows,
  As "Oh yes!" "Oh yes!" they cry.

"Listen to the Royal decree,
　　Whoe'er can whip her
　　Foot in the slipper,
She the Prince's bride shall be,
So says the Prince, says he!"

("*The Lost Chord.*"—SULLIVAN.)

For it may be his heart's bright angel
　　May walk in that shoe again,
It may be from some window near heaven
　　She may hear those *Grand Old Men!*

(*Hornpipe.*)

But see what agitation
Thrills the female population,
When the Royal proclamation
Has been issued to the nation!
Every maid in silken stocking
At the Palace Gate is knocking,
All obediently flocking
　　To the Prince's call.
　　　　With smiling face
　　　　They take their place,
　　But soon their hearts are beatin',
　　　　For 't is all in vain
　　　　They strive and strain,
　　They cannot get their feet in,

And oh! flushed their cheek
And words they speak,
Which will not bear repeatin',
For whate'er they do,
That little shoe
Is much too small!

("*The Three Old Maids of Lee.*")

Then the sisters, one, two, three,
They come to the scratch with glee,
But one has a heel,
And the other has a toe,
And the third has a deal
Of both *de trop*!
So three old maids they must be!

("*The Midshipmite.*"—ADAMS.)

With a long, long pull,
And a strong, strong pull,
Gaily, girls, on make it go!
It is much too tight,
And in sorry plight,
They sigh wearily, girls, heigh ho!

(*"The Bluebells of Scotland."*)

But where, meantime where,
   Might our kitchen maiden be?
Why, she's standing at the area gate,
   To see what she can see!

(*Recitative.*)

And hark! oh, hark!
   It is the crier's bell!
And mark, oh, mark
   The words they tell!
" Listen to the Royal decree!
   She who can whip her
   Foot in the slipper,
  She the prince's bride shall be,
  So says the prince, says he."
Oh what bliss beyond revealing,
To her heart's now appealing!
Oh what ecstacy of feeling,
O'er her senses is stealing!

(*"The Girl I left Behind Me."*)

And that dear little shoe she longs to clutch,
   For of what does it remind her?
Why of course of the night she danced so much,
   With the prince she left behind her.

("*The Danube River.*"—H. AÏDE.)

She oft since then had watched a spoon,
　But never, loves! Oh never, never
Had she observed so gone a coon
　Beneath her glances shiver,
Or been so near her honeymoon
　Somewhere upon the river.

(*Recitative.*)

Her head is dizzy,
　Tho' the prospect is so charmin';
And love is *busy*,
　Her sweet spirit *calmin'*—
　　　Bizet!
　　　Carmen!
Thus my pun I poke,
Surely you'll observe my little joke!

("*Habanera.*"—CARMEN.)

Love is whispering in her ear,
　"For you an ex'llent marriage this, my dear;
To have ever so much a year,
　And to live in a most exalted sphere!

Lots of palaces everywhere!
  Genuine diamonds and fictitious hair!
By Elise to be dressed, and ne'er
  The self-same dress more than once or twice to wear!
        'Tis Love, 'tis Love!
For Love is not a god so blind
  But what he 's wide awake enough sometimes,
And in this world 'tis odd we find
  How often interest with passion chimes!
Says Love " Compete, dear maid,
  This matrimonial slipper try!
Just show your feet, dear maid,
  And let the monarch mind his eye! "

("*Men of Harlech.*")

  Full of such unselfish feeling,
  Depth of character revealing,
  See the maiden forth is stealing,
  To the palace gone!
        Sits her down
          So lightly!
        Lifts her gown
          So slightly!
  And the sisters, seeing what she 's after,
  Shake each Royal beam and rafter,
  With their indecorous laughter!

5 *

Ha! ha! ha! sounds through the palace,
Soon at an end their mirth and malice,
With her foot the slipper tallies,
    And she draws it on.

("*He's got 'em on.*")

    She's got 'em on!
    She's got 'em on!
One upon each tiny tootum;
    She's got 'em on!
    She's got 'em on!
Oh, and don't those slippers suit 'em?
With wonderful dexterity she's got the whole pair of 'em on!
    And all the Court with bows salute 'em,
        They declare
        She's all there,
Because she's got those pretty slippers on!

("*German Volkslied.*")

    See! he is by her now,
    Joy on his princely brow;
    Round her his arms he throws;
    Sisters blush, and cry Oh, Ohs!
    And they were marriéd,
    Yes, in church were wed,
    And the tears were shed,

And the guests were fed,
And nice lives they led,
  'Till they both were dead—
And that's as far as we need go!

("*Wedding March*" (2nd *Part*).—MENDELSSOHN.)

Now, if my moral you'd not be losing,
This it is—in sponsors choosing,
  Mind you for good fairies look,
In which case you safely may wed your cook!
  Now, is not my tale a ripper?
Loveliness with virtue blended,
Ugliness by vice attended,
      All splendid!
      All now ended!
  Cinderella and her small glass slipper!

BLUE BEARD.

BLUE BEARD.

("The Hollybush Polka.")

I will sing you a deadly
Form of song called a medley,
You can sleep through it stead'ly
    If it strikes you as rot ;
But if you think, when you waken,
That it's Shakespeare or Bacon,
You'll be very much mistaken,
    For it's not!

("Ein Jüngling liebte eine Mädchen."—M. V. WHITE.)

Now the hero of my poem
    I first must disclose to view,
All Britons from babyhood know him,
    And so do grown-up babes too,
        Yes, you do !
For he's named from his beard which was blue,
        There's a clue !

("*The King of Mashers.*")

He never used lotion,
    He never used dye,
He hadn't a notion
    He looked such a guy!
As blue as the ocean,
    Or tropical sky,
He never knew wherefore,
    He couldn't tell why!
Such an oddish-and-weirditty, much-to-be-feareditty,
    Blueditty-Bearditty swell ah!
Such a hankitty-pankitty, quippy-and-crankitty,
    Mooditty-Sankitty man!
So sad he and serious, mad and mysterious,
    Bad and imperious fellah,
Of the don't-play-the-fooleous, strictly Home-Ruleous,
    Cæsary-Julius clan!

("*Ein Jüngling liebte eine Mädchen.*"—M. V. WHITE.)

He was very rich and designing,
    And a nice little maid he did woo;
She had no idea of declining,
    There was something so natty and new
        In the hue,
Of a man with a beard which was blue,
    Through and through!

("*That was All.*")

He met her but once in the Soho bazaar,
      That's all!
She was shopping with sweet sister Anne and Mama,
      That's all!
He winked, and she said, "You are going too far!"
He murmured, "Oh, be my particular star,
My Harem's fair Houri!" She shouted, "Hoorah!"
      That's all!

("*Stéphanie Gavotte.*"—CZIBULKA.)

But she'd another and an earlier flame,
  She loved and she was loved by him,
I cannot recollect his Christian name,
  It isn't given in my Grimm;
Ah! he was beautiful, and good, and brave,
  All that a young knight ought to be,
But rather hampered by the troublous wave
  Of impecuniosity!
Many were the walks they had together,
Many were the talks they had together,
Many were the vows, the "love-me-then-as-nows,"
And other little rows they had together!
Pretty little birdies of a feather,
Life for them was all sunshiny weather,

Always bill and coo, as youthful lovers do,
  There's nothing very new in *that*!
But what was shocking was the way in which,
  When rank and fortune came in view,
She to the poor young man preferred the rich,
  And promptly married Mr. Blue—
                    Blue-Beard!

("*Tut, Tut.*")

Tut! tut! tut! tut!
  Who'd a thought it?
Tut! tut! tut!
  You don't say so!
Tut! tut! tut! tut!
  Who'd a thought it?
Goodness me! what a fearful go!
Goodness me! Can a girl do so?

("*The Bailiff's Daughter of Islington.*")

And with that pair
Dwelt a spinster rare,
Who is part of my story's plan;
She was much more blue
Than the other two,
And they spoke of her as "Sister Anne."

("Vergiss mein Nicht.")

Sweet sister Annie!
She'd never smile nor flirt,
She spoke of man as dirt,
   Poor grovelling man!
She'd woman's rights assert.
Waist never tightly girt,
Wearing divided skirt,
   Sweet Sister Anne!

("As long as the World goes round.")

As long as the world goes round,
Such strong-minded maids are found;
   First happy and careless,
   Then cappy and hairless,
As long as the world goes round!

("O Dainty Triolet."—PRINCESS IDA.)

   O happy triolet,
   Of blue or violet,
   Who thus unitedly
   Could live delightedly;
   O sweet society
   Of worth and piety,
Where life satiety could never bring!

O home so quiety,
Which with propriety
Its ululiety of joy could sing,
With nice variety
Of drink and diety,
Where inebriety was not the thing.

("*Couplets de la Rosière.*"—BARBE BLEUE.)
Said Blue-Beard to his wife one morning,
"My love, I've had a little warning
              To go!
The fruity port my nose that tinges
Has given me some nasty twinges
              In toe!
My doctors in consultation all
     On German waters are agreed,
So quick now a hansom call,
     And mind you choose a lively steed;
And if you ask me where I'm going,
     I say, 'To try some hydropathic cures';
And if you further would be knowing
     I simply add, 'that's no affair of yours!'

("*Simon the Cellarer.*")
"But I leave you (they call this the confidence trick)
     The keys of my boxes and doors,
You can taste every vintage and make yourself sick
     With pickles and jams from the stores,
          The Armico-nautical stores!

But there's a small cupboard behind the back stair,
   And, Fatima, mind you are *not* to go there!
On your freedom this single restriction I leave,
   Don't be *dead* on that *key*, but be on the *qui vive!*"
But "Ho! Ho! Ho!" she protests with much go,
   "What! disobey Blue-Beard? O dear me, no!"
And Ho! Ho! Ho! How her promises flow,
   "What! disobey ducky dear? O dear no!"

(Recitative.)

But 'ere he leaves his pretty one all widowed and
   alone,
He'll give her one short ditty with his matchless
   baritone!

(*" Maid of Athens."*—GOUNOD.)

  "By those tresses varying tinge,
    By that false and fanciful fringe,
    By thy sad and Santley cheek,
    And by thy taste for modern Greek;
    By the music of Gounod,
    By three-acred cows that low,
    Fatty *moo*, sas agapo!"
    What that means you probably know!

(" *Excelsior.*"—Miss LINDSAY.)

The shady knight was going fast
When thus her clarion answer passed,
And tho' her notes were far from right,
Her voice was at a fearful height—

(" *Good-bye,*"—TOSTI.)

FAT. " Good-bye! my husband, good-bye! good-bye!"
ANNE. " Good-bye! my brother-in-law, good-bye! good-
bye!"

(" *Lieder ohne Worte.*"—1er *Recueil*, *No.* 4, MENDELSSOHN.)

Then she hears her youthful lord,
Her *leader only forty*,
As he canters and he gallops,
With a clitter and a clatter
And a pitter and a patter,
Fainter growing,
Further going
From his home,
For he is going from his home, home, home!
What will she do? Will she be circumspect now?
Oh, dear me, no! she'll to herself be true;
And if you understand the female sect now,
You'll know at once what she is bound to do.

("*The Night is Dark.*"—RIP VAN WINKLE.)

Oh, fatal female failing,
Over prudent thoughts prevailing,
Direst consequence entailing
Curiositee!
Forgetting lord and master,
But with steps now fast and faster,
Rushing headlong to disaster
Fair Fatima see!

("*Lieder ohne Worte.*"—*No.* 4, continued.)

Now she is standing
On the fatal landing,
To the secret chamber speedy entrance demanding!
Will she unlock it?
Yes, from her pocket
Draws the little key that fits in the socket!
One turn! 'Tis done! the door flies open wide!
And ah! ah! ah! what horrid sight she sees inside!
And now again she hears
Her *leader only forty*,
As he canters and he gallops,
With a pitter and a patter
And a clitter and a clatter,
Growing clearer,
Coming nearer
To his home
For he's returning to his home, home, home!

(Recitative.)

But what on earth can those forms be
That glare on her so fearfully—
Those ghastly figures ranged in rows
All stiff and stark in death's repose?
She screams! she starts! she drops her key!
Ah! goodness me! why these must be

("Three blind Mice.")

Ten dead wives! ten dead wives!
    See how they stare!
By halter, by poison, by carving knife,
Each one of these ladies has lost her life,
Just simply because she was Blue-Beard's wife,
    Ten dead wives!

("Ten little Niggers.")

The first had a squint, and the second had a cast,
The third was too slow, and the fourth was much too
    fast;
The fifth of that batch was decidedly the worst,
And they'd all looked in the cupboard, except of course
    the first.
The sixth was a flirt, and the seventh was a prude;
The eighth and the ninth were not exactly what they
    should,

The tenth had gone much further than the other nine,
  but ah!
When they looked in the cupboard, she wasn't *gone* so
  far!

("*Légende de la Langouste*."—L'ŒUIL CRÉVÉ.)

Then she tries the key to clean
With some fragrant benzoline;
Rubs with turpentine and soda,
Horrid fears and fancies goad her,
"Out," she cries, "Oh, damnéd spot!"
—Sounds bad language, but it's not—
Wears out fifty scrubbing brushes,
Still the key ensanguined blushes;
Tries some pails of Pears's soap,
When *that* fails, oh! dead is hope!
          Now he is there,
You can hear his spurs a-clinkin',
          Yes! you may swear
He'll be with her in a winkin',
          Hear how he knocks!
          Great ugly shocks!
          Bang! bang!
      Rat-atatt-atatt-atatt!
          Now on his way
Up the staircase he is tearin'!
          What does he say?
Oh! I think he must be swearin'!
Oh, what a din! Oh, what a din!
      "Come in! come in!"

("O Willie, We have missed You.")

Fat.	Oh, husband, is it you, dear ?
Blue.	Yes, yes, it's me !
Fat.	You're looking rather blue, dear !
Blue.	Come, give me back my key !
Fat.	I heard you at the door,
	And I knew your welcome sneeze—
Blue.	My good woman, don't you bore,
	But give me back my bunch of keys !
Fat.	Here they are, love, all correct ;
Blue.	What ! the little one is specked !

(Recitative.)

Blue.   Oh, Fatima, what do I see ?
A mark upon my favourite key ?
Now, come, you know it was not so,
When I left you here an hour ago ;
    It wasn't there,
    I dare to swear,
When I went away on my foreign visit ;
    So kindly deign
    At once to explain
What is it, woman, what is it, what is it !
    Explain, woman—plain woman !

("*Now to the Banquet we press*" (*Last Part*).—THE SORCERER.)

FAT.	The mark on the key, love,
	You 're asking of me, love—
	The mark on the key, love,
	You see, love—you see, love—
	I cannot tell what,
	But I think it 's a spot;
	Yes, I take my Sam,
	It really is not
	What you think, but a clot
	Of good strawberry jam!
BLUE.	Jam?
FAT.	Yes, it 's currant and raspberry jam!
BLUE.	Cram!
FAT.	Cram?
BLUE.	What a horrid, detestable cram!

("*Dopo.*"—TOSTI.)

His face grew black, and the eyes were cold and frosty,
   With which his trembling spouse he transfixed;
His hair was tumbled—in fact, I might say *tossed*—he
   Addressed her in language that was rough and mixed,
"O Fatima, perché, warum, pourquoi avez-vous osi
Nel sepulchro dei miei poveri sposi
Stickiare tuo impertinente nosi?
Infatuata, was hast du gethan?
Du und deine schwester Ann?"

("Di quella Pira."—TROVATORE.)

" Now then, my Fatima, prepare!
  For that snug cupboard you must share
  With those your predecessors there,
     My wives deceased!
  Take down your interesting hair,
  But leave your neck a little bare,
  For I should like to hit you there,
     I'm such a beast!"

(Recitative.)

FAT.   O Pieta, pieta! Orribile!
BLUE.   No, Fatima, no, Fatima! Impossibile!

("Father, come Home.")

FAT. Oh, Annie, dear Annie, come up to me now!
  The clock in the steeple strikes three!
And I fear that, before it has struck many more,
  Blue-Beard will be striking of me!
'T is true I ne'er loved thee, sweet sister, till now,
  But fondness is fostered by fears,
And when pain and when sickness are racking the
    brow,
  An angel my Annie appears!
      Come up! Come up! Come up!
      Oh, Annie, dear Annie, come up!

("*As Usual.*")

Now she was studying down below.
        As usual !
Her Aristotle and Plato,
        As usual !
And writing pamphlets just to show
That men are useless drones and low,
And that without them things would go
        As usual !

("*All the Afternoon.*")

Then an answer quick she passed her,
With a face like mustard plaster,
" I of Arts shall ne'er be master,
    If I leave my books so soon ;
For the other girls from Girton
Will be beating me, that 's certain,
If I cannot keep the spurt on
        All the afternoon ! "

FAT.   What ! all the afternoon ?

ANNE.   Yes ! all the afternoon !

But as you seem so sorely prest
    I 'll put my work away,
And take the thing I most detest—
    A British holiday !

("*Tom Bowling.*")

For although you to me have not been very jolly,
    Oh many a time and oft!
Yet, as you now seem so melancholy,
    Why, of course, I'll come aloft,
At once I will proceed to come aloft!

("*Nelly Bly.*")

FAT.   Annie, my
          Sister spry,
Mount the old clock tower,
          For wretched I
          Have got to die
In quarter of an hour!
Just take this little opera glass,
    And keep a good look out,
And if you see a Bobby pass
    Just give the man a shout!
Hey! Annie, Ho! Annie, tell me what you see,
Bawl to me! squall to me! and blow the melody!

("*Arabesque.*"—SCHUMANN.)

FAT.   O Sister Anne, O Sister Anne!
       What do you see? What do you see?
ANNE.   I see a man, I see a man!

Who can he be?   Oh goodness me!
It is a tree!
Oh misery!
It is an ordinary tree!

("*Wächterlied.*"—GRIEG.)

BLUE. Ten minutes done!
The fatal hour is drawing nigh!
Five more to run!
Prepare, O Fatima, to die!

FAT. Ah! your forgiveness I implore!
I'll never be inquisitive any more!

BLUE. Too late! too late your compunction!
 Spare yourself fatigue!
For I shall slay you with unction,
 Merry as a *Grig!*

("*Be kind to the Dear Ones at Home.*"—CHRISTY.)

FAT. O Blue-Beard be kind
To the pets left behind,
My poodle and pussies that mew,
And if in the corn
You should hear his small horn,
Be kind to my little boy blue!
Be kind to my sister, the learned Annette,
Excusing the length of her jaw;
Be kind to my mother, and strive to forget
That she is thy mother-in-law!

(Recitative.)

ANNE.	Oh I see! Oh I see!
FAT.	What? What?
ANNE.	Oh I see such a lot, such a lot!

(Sailor's Hornpipe.)

Oh I see some sheep a-feeding where the daisies bloom,
And the profile of a pig upon the sky-line loom;
    I see a swooping eagle
    Which is trying to inveigle
    A poor idiotic seagull
      To its doom, doom, doom!

("That's All.")

And now on the breezes is wafted a bray,
      Hee haw!
'T is the biped that much prefers thistles to hay,
      Hee haw!
And the sounds in her heart softest memories woke,
The voices of friends and the words that they spoke,
And again in her ears sings that affable moke,
      Hee haw!

("*What a Bit of Luck!*")

For the knight that she loved had been wonderfully true,
    What a bit of luck!
And the sister on the tower-top was doing all she knew,
    What a bit of luck!
She was banging with a banjo, she was thumping with a
    tong,
She was whooping with a whistle, she was grinding on
    a gong,
And the young man he heard her as he rode along,
    What a bit of luck!

("*Only a Song.*"—DE LARA.)

    Only a gong!
Its voice was loud, so loud and cracked,
    And he who heard was lean and spare!
His hungry lips the young man smacked,
    And smelt roast mutton in the air!
So thought he thus by fancy buoyed,
    And longed each dainty dish to munch,
He said unto his aching void
    "It is the gong! 'T is lunch!"

("*On this Subject we pray You be Dumb.*"—MIKADO.)

FAT.   Now come, sister Anne, you must something see,
      For my minutes of life are reduced to three!
         And Blue-Beard, fury breathing,
    His sword is unsheathing, his sword is unsheath-
         ing,
         And soon it will be seething
         In this little wee thing,
           This poor little me!
           In view of this crime
         I'm sure you'll agree,
          It's really high time
    You something should see, you should see!

(*Recitative.*)

ANNE.   Oh yes! I know I really *must*,
      For I occupy a post of considerable trust;
      And even now a cloud I spy,
      And there's more in that cloud than meets the eye;
      A cloud mysterious, ill defined,
      But I bet you a monkey that there's something
        behind!

(*" Wait till the Clouds roll by."*)

Yes! yes! in the distant vista
    I something substantial spy!
But I think 't would be best, sweet sister,
    To wait till the clouds roll by!

(*" The Skipper and his Boy."*—H. Aïde.)

She 's watching, still watching, when " Ho ! " cries she,
" I see a brave rider come fast and free ! "
Her long raven tresses she waves in her hand,
They 'd been tied to her head by a velvet band;
And all that doth of that head-dress remain
Are the hair-pins she 'll probably use again;
But she knows that the knight doth her signal see,
As they ride together, his donkey and he,
As they ride together, his ass and he.

(*"Good-night."*—*Plantation Song.*—A. S. Gatty.)

ANNE.   Fast! youth, fast!
BLUE.   Time is up, young woman, come!
ANNE.   Blast! youth, blast!
         (Rumti-tumti-tumti-tum!)
         Blue-Beard calls!
BLUE.   Oh yes, I call!
FAT.   This moment is my last,
            Then, good-night, sister Anne, that 's all!

("*Dream Faces.*"—HUTCHISON.)

Youth most attractive, rushing up the stairs!
Youth strong and active, taking steps in pairs!
Whistling so gently still the old refrain,
Hope on, dear loved one, here we are again!

("*The wearin' of the Green.*")

Now when Blue-Beard saw his rival shaking hands with
    sister Anne,
And advancing towards Fatima, with some much bolder
    plan,
He was very much put out, indeed, and didn't know what
    to do,
For the presence of that nice young man was wearin'
    to the Blue!

("*In Passion's Trance.*"—M. V. WHITE.)

With passion's trance his face was *white*,
   A knight of *maudlin colour he*!
He didn't wish, he didn't wish to fight,
   His way to flight he could not see!
The young man stood all at her side,
With tenderness her face he eyed,
And asked her thus to be his bride.

(*"Pretty little Darling."*)

"O! you little darling, I love you!
  O! you little darling, sweet Mrs. Blue!
If you really love me as you ought to do,
Blue Beard I'll at once proceed to carve in two!"

(*"Two lovely Black Eyes."*)

Then at him he flies!
  Oh! what a surprise!
Only for telling your wife she was wrong—

(*"The Bay of Biscay."*)

Then loud the battle thunder'd,
  And the blows in deluge poured,
Till wounds about one hunderd,
  Old Blue-Beard's rival scored;
At the end of the fifteenth day
  The monster lifeless lay!
Thenceforth away, happy were they,
  All on the gay, and the frisky, Oh!

(*"Only an Orange Blossom."*)

Only an orange blossom,
  Only a wedding cake,
Only a single bridesmaid,
  Which part sister Anne must take.

Only a happy marriage,
　With a breakfast light and gay,
Only a nice open carriage,
　To bear the pair away!

("*Vittoria.*"—CARISSIMI.)

Victoria! Victoria!
　They start in a trice!
　There's nothing so nice
　For couples that splice
As the well-built Victoria,
　Which Hooper supplies!
　　Mid slippers,
　Old slippers and rice!

("*I've got to do without 'em.*")

Now my moral, of course, for wives is that they must
　not be so curious,
　Or if they must, they should not trust old men with
　　beards of blue;
And for husbands, that wife-murder to themselves may
　prove injurious;
　Now, if more morals you require, I'll tell you what to
　　do—

You 'll have to do without 'em,
 You won't have any more !
You 'll have to do without 'em,
 Just as lots have done before.
So, good-bye happy couple !
 And farewell sister Anne !
  And adieu to you,
  You beard so blue !
Oh, good-bye, you grand old man !

SONGS.

THE BAILIFF'S DAUGHTER OF ISLINGTON (FRENCH VERSION).

Il y avait un garçon,
　　Bien aimé de tous les gens,
Et il était le fils du Lord Maire,
　　Et il aimait la fille
　　Du Sergent-de-ville,
Qui demeurait á Leycesster Sqvare.

　　Mais elle était un peu prude,
　　Et n'avait pas l'habitude
De coqueter, comme les autres demoiselles,
　　Jusqu'à ce que le Lord Maire,
　　Homme brutal, comme tous les pères,
L'éloigna de sa tourterelle.

　　Après quelques ans d'absence
　　Pour le chercher elle s'élance,
Elle se fait une toilette de tres bon goût!
　　Des pantoufles sur les pieds,
　　Des lunettes sur le nez,
Et un collier sur le cou—c'était tout!

Mais bientôt elle s'assit
Dans la rue Piccadilli,
Car il faisait extrêmement chaud;
Et la elle vit s'avancer
L'unique objet de ses pensées
Sur le plus magnifique de chevaux!

" Je suis pauvre et sans resource!
Prête, prête-moi ta bourse,
Ou ta montre pour me montrer confiance?"
" Jeune femme, je ne vous connais,
Ainsi il faut me donner
Une addresse et quelques références!"

" Mon addresse—c'est Leycesster Sqvare,
Et pour référence j'espère
Que la statue du Shakespear vous suffira!"
" Ah! connais-tu na mie,
La fille du sergent?" " Si;
Mais elle est morte comme un rat!"

" Si défunte est ma belle,
Prenez, s'il vous plait, ma selle,
Et ma bride et mon cheval incomparable,
Car il me faut rien dire,
Mais vite, vite m'ensevelir
Dans un désert sec et désagréable!"

" Ah! mon brave, arrête-toi!
Je suis ton unique choix,

La fille du sergent sans peur !
        Pour mon trousseau—c'est modeste,
        Vous le voyez ! Pour le reste
Je t'épouse dans une demi-heure !"

        Mais le jeune homme épouvanté
        Sur son cheval vite remontait,
La liberté lui était trop chère !
        Et la pauvre fille dégoutée
        N'avait qu'a reprendre sa route et
Se retourner chez monsieur son père !

    .   .   .   .

Elle n'a jamais depuis vu le Lord Maire !

    .   .   .   .

Elle s'addresse encore Leycesster Sqvare !

THREE FISHERS (FRENCH VERSION).

———o———

Trois pêcheurs se mirent dans un p'tit bateau,
Pour attraper les poissons qui nagent dans l'eau,
Enchantés de quitter leurs enfants si chers
Et leur femmes, qui grondaient bien plus que la mer !
  Car les hommes travaillent,
  Et les femmes chamaillent,
 Et le plutôt c'est fini le plutôt sommeil,
  Malgré le son dégoûtant
   De l'ouragan !

Trois femmes au cinquième, sans maris, très gaies !
Qui se disaient des scandales en buvant leur thé ;
Dont chacune au même instant parlait très haut,
Et claquait ses enfants—une vingtaine ou so !
  Mais que voulez-vous donc ?
  C'est toujours le même chant !
 Beaucoup de bébés, et point de savon !
  Malgré les vagues et les vents,
   Qui font concon !

Trois cadavres horribles sur le sable ! O malheur !
De ceux qui avaient été les trois pécheurs !
Et trois femmes, qui jetaient leurs chignons dans l'air
De ce que n'avait jamais été leur " hair " !
    Car les hommes se grisent,
    Et les femmes se frisent,
Et rien est affligeant quand on est bien mise !
    Adieu ! pères, mères, enfants !
      Et les éléments !

EXTRACT FROM "SOMNAMBULA" BURLESQUE.

———o———

("*Let's give three cheers for the Sailor's bride.*"—PINAFORE.)

CHORUS OF VILLAGERS.

Now here's three cheers for the blushing bride,
The prettiest girl in the country-side!
For of all the maids in the neighbourhood
There's none that's so pretty and none so good!
With a tra-la-la-la!
And a fal-a-lal-a!
And whatever's the Italian for hip-hip-hurrah!
With a shout and a holloa,
And noises to follow
Expressive of mirth, Ha! Ha!
Now here's, &c.

(*Recitative.*)

AMINA (*at window*).

O amici miei cari,
Compagni della mia giovinezza,
I am about to marry,
And I thank you for your dolce politezza !
I was taking a riposa,
Just a little kind of doza,
A siesta, half undresta,
When I heard your serenata,
Che il fenestro penetrata
Del privato appartimento,
And I felt, mi sento grata
    E contento !

'T is true I'm del villagio la bella,
La rosa, la raggia, la stella !
But that which would puff up another fellah
Don't make *me* a conceited damigella ;
For I know that from the ciel
Comes the beauty of this girl,
This fair face and these bel occhio
Which make men talky-talky-oh,
And that if I'm a Venus,
As say plenty of the genti,
All that *I've* done 's been to add clothes,
    Vestimenti !

My style of singing's pure Italiano,
Mia voce's a magnificent soprano ;
Not lost (perdute) is its beauty, Ah no !
I still have half the compass of a Piano !
But though dolce far niente
Are my notes—and I have plenty,
I do not disdain, amici,
Your cantando crude and screechy.
For I know 't is for my glory
That you sing so con amore,
And it melts il cor deep in mio
                    Interiore !

For hard would be my heart and stony Oh !
Could it contemplate unmoved my matrimonio !
This is the proudest moment I e'er spent Oh !
Della vita superbissimo momento !
Unaccustomed as I am to
Make a publico discorso,
I can *not* resist this oppor-
tunity to sing a morceau,
A small melody Bellini wrote,
And christened " Vi Ravviso,"
And " How ravishing," you 'll say, cos'
                    It suits me so !

("*Vi Ravviso.*")

"Vi Ravviso's" a cavatina
  Well adapted to this Amina.
  For it's easy, not hard upon her,
  And she's wheezy—this little Prima Donna—
      And it expresses,
      (As do her dresses,)
    That she is young and gay,
      As bird in May, as bird in May!
  With joy my little heart is bounding,
  With my rapture the village is resounding.
      List awhile!
      You'll hear me smile,—
  Ha, Ha! Hee, Hee! Hoo, Hoo! Ho, Ho!
      For I have captured
      A youth enraptured,
  And he's a good parti, hee! hee!
  Yes, he's a good parti;
      I have made a lucky hit,
      And I am very glad of it.

SCHUMANN'S NOVELLETTE.

"Go, John, go, the doctor bring,
For Master Tommy's not the thing!
All night long you might have heard him moaning,
Gnashing teeth, muttering, grunting, groaning!
Sounds like these betoken woe,
My child is ailing—go, John, go!"
Anxious heart—that of a widowed mother
With one son—when she's not got another!
Doctor comes! oh! great relief!
To whom she thus explains her grief.

DIE ÄNGSTLICHE MUTTER.

"Doctor, doctor, tell me quickly
    What doth ail my beauteous child?
Why he looks so sad, so sickly,
    Cheek so pale and eye so wild?
Speak, oh! speak, and tell me true, man!
Sprich doch, lieber Doctor Schumann!
This to know I must insist
Was mit ihm der matter ist!
Is it croup or scarlatina,
    Measles, mumps, or fever low?

Why so changed his whole demeanour?
Why his cheeks grow green and greener?
    Must he to the angels go
    Ere his wings have time to grow?"

DER GESCHICHTER HERR DOCTOR.

"Sei du still, ich geschwind
    Examiniren muss das Kind:
    Öffnet jetst ihr kleines mund, so!
    Och! ihr zunge ist ungesund so,
    Gieb die hand—Och tier, Och tier!
    Das puls ist hundert vierzig vier!"

DAS KRANKES KIND.

        "Oh! Oh! Oh!
        It comes again,
        That horrid pain!
        Ah me! Was ein schmerz
                Am herz!
        It there must be,
    Or somewhere in that locality!"

DER DOCTOR.

    "Ja wohl! ja wohl! mein kind,
    Du bist, du bist sehr krank;
    Was hast gegessen
    Und getrank?"

DAS KIND.

    "I last night was at the table
        Table spread with Christmas cheer,
    Eating all that I was able—

Christmas comes but once a year!
Turtle soup so green and fatty,
Lobster salad, oyster patty,
Champagne, hock, and bottled porter,
(And of each more than I ought 'er):
Roast beef, turkey, stuffed with truffles,—
 Sacred Christmas duties these—
Plum pudding in flames of brandy,
Six mince pies, which came quite handy,
 And on top I chanced to squeeze
 Two stout layers of toasted cheese!"

DER DOCTOR.

 " Donner und blitz! mein sohn,
 Das war zu viel, zu viel
 Für einem kind
 In einem meal!
Jetst du must trinken das oil von castor,
Und am brust stricken ein mustard pflaster!"

 Morgen kommt! Ganz wohl ist er!
 Der Tommy ist sich selbst einst mehr!

Now the moral is to eat just enough, not too much—
Nur genug, nicht zu viel, ist das moral in Dutch—
And this was Schumann's meaning, well I know,
For Madame Schumann herself told me so,
 She herself told me so,
  She told me so,
   So! so!

London: Printed by W. H. Allen & Co., 13 Waterloo Place. S.W.

www.ingramcontent.com/pod-product-compliance
Lightning Source LLC
Chambersburg PA
CBHW022140020726
47496CB00008B/2490